WARM-UP PREVIEW

Operating under the assumption that you are already well warmed-up, we ask you simply to read this brief introduction.

The last unit was an introduction to some of the knowledge needed by the engineer as a practitioner. This unit is an introduction to some of the knowledge that he or she needs as a professional. It is concerned mainly with the nature and significance of the engineer's interaction with society. Engineers do not operate in a vacuum. Far from it. Their jobs, attitudes, and contributions are strongly affected by the economic, organizational, social, political, and legal environments in which they operate. Engineers, in turn, primarily through their creations, significantly affect employers, consumers, the public at large, and a variety of economic, social, and political institutions. This multifaceted interaction is the subject of the next three learning segments.

Learning Segment 13
THE IMPACT OF ENGINEERING

LEARNING OBJECTIVES

- Understand that engineering has an impact both at the level of the specific technological venture and at the level of the collective effect of many such ventures or of all technology.

- Recognize some of the problems that engineers have with second-order, or indirect, effects of engineering creations.

- Recognize that engineers must not only know the human wants and needs in an engineering venture, but must also actively show they care about those needs and wants.

- Develop a sense of the obligations arising from the social impact of engineering.

INTRODUCTION

Tools, machines, and structures have had far-reaching effects on mankind; the evolution of these contrivances has been closely intertwined with major political, social, military, and economic events throughout history. Visualize, for example: the commercial, cultural, and political interactions between nations that resulted when vessels capable of traveling the high seas were developed; the increase in agricultural productivity when the iron plow replaced its wooden predecessor; and the impact of the printing press on the preservation and dissemination of knowledge.

The creations of modern engineers are no different in this respect. The automobile obviously has significant social, economic, and environmental effects. Agricultural mechanization has greatly hastened our transition to a predominantly urban society. Mammoth power plants convert fuels into energy that, in turn, underlies our industrial and agricultural might; this, in turn, has enabled us to become a society with an unprecedented standard of living. And is it necessary to mention "the bomb" and its impact on the affairs of men? Or automation? Or the computer?

This learning segment explores the impact of engineering at two levels. The first, the obvious, and the simplest to explain involves the effects of an individual creation (a giant machine for irrigating the desert, the snowmobile, or an emission control device for automobiles). The second level of impact involves the *collective* effects of many creations, generally a certain class of them (the aggregative effects of modern communication media, the total effect of all machines on the atmosphere, or the impact of all appliances and other machines on family life). This type of impact is more elusive; it is generally difficult to

MODERN ENGINEERING

Engineering and Society

Edward V. Krick

Lafayette College
Easton, Pennsylvania

Wiley Professional Development Programs

Advisory Editor
Steven C. Wheelwright
Harvard Business School

John Wiley & Sons, Inc.
New York • London • Sydney • Toronto

Library of Congress Catalogue Card Number: 76-10038

ISBN 0-471-01702-7

Printed in the United States of America.

10 9 8 7 6 5 4 3 2 1

perceive, let alone to influence. Consequently, the impact of the collective works of engineers will receive proportionately more attention.

IMPACT OF A SPECIFIC TECHNICAL VENTURE

Consider a familiar, relatively straightforward case: the construction of a new freeway in a suburban area.

THINK IT THROUGH

Take the case of a new superhighway to be constructed in a suburban area. You know that it has both benefits and costs. List two of each.

____

____

____

____

____

____

Some answers to this exercise appear in the following paragraph.

You know that it visibly affects the landscape, that it means added convenience (and safety, by the way) for millions of drivers, that it is noisy and smelly, that it displaces many families and disrupts long-established neighborhoods, that it influences land use for miles around for a long time to come, that it affects established drainage patterns, and that it drastically alters traffic flows on connecting roads and nearby streets. Surely, if you were locating and designing this freeway, you could not escape an awareness of these and other effects of your creation.

However, for every case like this in which the far-ranging effects are relatively apparent, there are many technical ventures for which the full range of effects is obscure. This is true in the case of nuclear energy, the plan to redistribute Arctic waters over North America, satellite transmission of television over a multinational territory, aerosol sprays, or an offshore tanker terminal.

The snowmobile might strike you as a relatively benign development and, as far as the United States is concerned, this may prove to be so, although some naturalists would surely disagree. However, in areas further north, the effects go far and deep.

THINK IT THROUGH

Can you think of a few effects that the snowmobile might have in such places as northern Canada and the Scandinavian countries?

____

____

____

____

____

Some answers to this exercise appear in the following paragraph.

In northern Canada and especially in northern Scandinavia, where reindeer herding is a primary industry, the arrival of snowmobiles has meant, in addition to the environmental effects:

- A significant increase in social interaction.
- Better medical aid and other services.
- A new industry associated with the sale and servicing of these machines.
- Doom for beasts of burden; sled dogs and reindeer are becoming obsolete.
- Larger harvests by fishermen, hunters, and trappers. The quantum jump in mobility provided by the snowmobile resulted in a virtual elimination of bears in one season in Finland.
- A major social problem that closely parallels what extensive mechanization has meant to small farmers in this country. The deer-herding industry is shifting from labor-intensive to capital-intensive, spelling economic doom for a majority of the small, marginal herders.
- Increased reliance on an external source of energy: oil.

Those Pesky Indirect Effects

Note that many of the consequences of engineers' creations are indirect, such as the socioeconomic changes precipitated by the snowmobile. Such consequences are often referred to as side effects, or second-order effects.

YOU DECIDE

Why would it be difficult to include second-order effects in a benefit-cost analysis of a proposed engineering project?

An answer to this exercise appears in the following paragraph.

Ordinarily, second-order effects are unintended; often they are unanticipated. On occasion they amount to a chain reaction of interrelated effects that would be virtually impossible to predict. Frequently the victims are "innocent third parties" who are neither instigators nor beneficiaries of the change. Television may have some adverse social effects, but the people who are thus affected are the same ones who use and enjoy the medium. But many of the persons who are displaced by new freeway construction are not beneficiaries of the new road; the same applies to factory workers who are displaced by new production equipment and office employees whose jobs are taken over by computers; and, too, many people who live near major airports and suffer the familiar adverse effects have never been in a plane. True, not all side effects are negative; there are second-order benefits as well as costs. However, indirect benefits of a technical venture, anticipated or not, are hardly cause for concern.

Second-Order Capabilities

Observe how engineered devices, machines, and structures underlie many of man's other important capabilities. Agriculture is an excellent case in point. It contributes what it does and employs relatively few people in doing so, mainly because of the variety and potency of the machines employed in most phases of the business (including in the large-scale manufacture of fertilizers and insecticides). The entertainment industry would scarcely have the impact it now has if it weren't for the widespread availability of motion picture machines and television facilities. Or consider medicine; without instruments and machines for diagnosis, surgery, and therapy, without artifical limbs and pacemakers, without mass production of drugs, how potent would medicine be? It is because of this "amplifying effect" of engineering creations in other areas of human endeavor that the term *technology* is so useful.

YOU DECIDE

The term *technology* poses some difficulties, since it means so many different things to different people. Before reading the following paragraph, try to write a brief definition or description of what technology is and what sorts of things it includes.

It may be impossible to get any consensus on the meaning of the term technology, but some thoughts are offered in the following paragraph.

The term *technology* embraces not only television facilities, but all of the capabilities of the entertainment-communication complex that derive from that equipment. This same is true for all of the implements, machines, mass-produced chemicals, and the like that engineering provides to agriculture and medicine, and everything these make possible in those fields, and for what engineering contributes to manufacturing, commercial fishing, food processing, dentistry, warfare, and so on. And, if you doubt that these capabilities are highly engineering-dependent, picture what remains when you take away the engineer's contrivances that underlie any one of those capabilities, such as medicine.

Observe, then, that the impact of engineering may be further reaching than you suspected, in view of the powers engineering gives to others which, in turn, have their own direct and indirect effects.

YOU DECIDE

Reflect for a few moments on the implications of the invention of the ordinary brick for people who previously relied on relatively crude tents for housing. What are some of these implications?

The obvious consequences are security and comfort (cooler in summer, warmer in winter). But the brick also had economic effects—starting an industry that has survived over 10,000 years. It also made villages more likely and enduring (there is quite a difference between the permanence of brick and the permanence of tent structures), and gave humans greater freedom of choice in where they could settle.

Enough, for the moment, on the impact of engineers' creations on human affairs and human welfare. This brief sampling of a story that fills many volumes is intended to make you "impact conscious," generally sensitive to the manifold ways in which engineers' creations profoundly affect our physical comfort and safety; our power to control; our capacity to injure; our mobility; the ease with which we communicate; the education we need; our life span; the hours, content, physical demands, and stability of our jobs; our environment; our values; and our behavior. The next step is to explore the consequences of all this for engineers.

THE CONSEQUENCE FOR ENGINEERS: SOCIAL RESPONSIBILITY

Engineering creations directly affect people, usually many people and in many ways. People ride transit systems, people operate machines in offices and factories, and people service automobiles. The obligation to serve well the people directly affected by their creations is an important professional responsibility for engineers. But it is not the only one; society is not going to let them escape some responsibility for adverse side effects like the social disruption of communities by urban freeways. Thus, engineers are expected to recognize the total physical, economic, psychological, and social impact of their solutions, including the indirect effects, the unintended consequences, the unquantifiable results, and the implications for nonusers.

If one is to fulfill these responsibilities, one must come to "know" the society affected and to care about its well-being. This "knowing" and "caring" require some explanation.

In all their projects engineers should learn what people need, what they prefer, and what they will tolerate. This awareness should be reflected in their solutions. If the engineer designing a new transit system for a community is to satisfy the maximum number of people, he should know their needs and wants.

THINK IT THROUGH

What are some of the needs and wants of people that would have to be considered in the design of a new transit system?

____

____

____

____

Some answers to this exercise appear in the following paragraph.

Where do people want to travel? When? How much importance do they attach to privacy? To pickup frequency? To comfort? To other criteria? Are enough people willing to give up the privacy and independence afforded by automobiles to make a new transit system feasible? How do they feel about overhead train-carrying structures passing through suburban areas? What noise level do they consider reasonable? What do they consider esthetically pleasing? This is all part of getting to know the people directly and indirectly affected by the system. While designing it, he should use this information to predict how people will be affected by and react to alternatives he is considering (for example, overhead versus background, individual cars versus trains) in an attempt to maximize satisfaction and minimize adverse effects and reactions.

Predictions of the economic, social, psychological, and political impact of technical alternatives require special insight concerning man's individual and social behavior. But knowing society is not sufficient; an engineer must also care. It is one thing for the designers of a large dam not to be oblivious of the psychological, social, and economic effects it will have on the farmers and villagers who will be dislocated. It's another thing to be aware and to be actively concerned about these effects by giving them due consideration in their benefit-cost deliberations, by minimizing the hardship created, and by facilitating the economic, social, and personal adjustments of the people affected.

LET'S BE SURE ABOUT THIS

What two things must an engineer have in relation to people's needs and wants?

____

____

____

An engineer must have knowledge of people's needs and wants. An engineer must also have care for people's needs and wants.

Engineers have been criticized for showing insufficient concern for the full implications of their creations. Some of this criticism is misdirected. Public officials and business executives make many of the decisions for which "the engineers" are unjustly blamed. However, while engineers are not to blame, they are not blameless; enough of this criticism is deserved to warrant some serious introspection.

True, engineers rarely have the ultimate say about what businesses and governments do with their engineering capabilities, but this hardly justifies not caring and not taking a tough stand on behalf of mankind. After all, we are as much creators of social change as of physical change; we should know and care about the victims as well as the beneficiaries of our creations. It is appropriate that our designs closely match human needs and wants, that we intelligently predict the full impact of our solutions, and that we minimize the adverse social effects and maximize the social benefits of our creations.

Few will disagree that engineers should learn about and care about the effects of their creations. However, what constitutes a proper professional response to the collective effects of engineers' creations–those described starting below–is not clear-cut. Yet, it would hardly be appropriate for engineers themselves to be ignorant of the consequences of their collective works.

COLLECTIVE EFFECTS

Some of technology's implications are direct and obvious. You have no difficulty noticing; you experience many of them. Then there are other implications, at least as important, that may well escape you, and often these are the collective effects introduced earlier.

For example, the effects of machines sometimes converge on a particular segment of the population. Look how machines–automobiles, trucks, farm equipment, and manufacturing machines–have combined to create an employment crisis for unskilled farm laborers. Agricultural machines, such as those that now pick 97 percent of the cotton crop, for example, have displaced thousands of laborers on southern farms. In fact, they have also put many operators of small farms out of business. In the meantime, some equally consequential developments are taking place in the urban centers. For a number of reasons, some of which are technological, factories are migrating from the city to outlying areas. This is feasible because automobiles have made employees so much more mobile and because trucks have cut the ties between factory and railroad. The result is that employment opportunities are shifting from city to suburb, during a period in which there is a major influx of job-hunters into the center city, many of whom lost their jobs to agricultural machines. They feel the pinch of machines again when they find that low-skill jobs are becoming scarce as a result of continuing mechanization in factories, offices, and construction.

Certainly you have observed the collective effect of machines on the family. For example, a small army of electro-mechanical slaves plus the automobile have made it much more feasible for both mothers and fathers to work outside the home. This, in turn, has acknowledged social and economic consequences.

YOU DECIDE

So says J. J. G. McCue: "Engineers seldom look upon themselves as inciters and fomenters of revolution. But a revolution need not be bloody; it is any drastic transformation of social institutions. For many years now, one of the principal sources of pressure on existing social

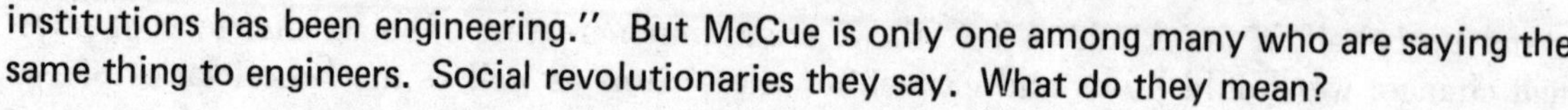

institutions has been engineering." But McCue is only one among many who are saying the same thing to engineers. Social revolutionaries they say. What do they mean?

____

____

____

The above-mentioned effects of automobiles and appliances on family life are impacts of technology you can readily relate to. These are not isolated cases. Note who it is that conceives and creates these contrivances: engineers. They cause more socio-economic change—through their factory and agricultural machines, transportation systems, and communication media alone—than people who hold social change as their main purpose in life. (True, engineers are social revolutionaries, but not deliberately, directly, or even consciously.)

Another, more subtle, illustration of the effects of technology is provided by the mass media. The printing press, radio, motion picture, and television have made their marks individually, but they also have their collective effects. One timely instance is their role in precipitating the recent surge in rising expectations of the world's poor. Awareness that men elsewhere have it much better is not the only factor in rising expectations; knowing of man's vastly increased power to accomplish has stimulated the world's poor to expect, and begin demanding, more.

Of course, modern communications media keep the more-privileged as well as the less-privileged better informed about what goes on in the world. This greater awareness probably contributes heavily to the diminished respect for certain institutions. Observe how much of what we are told by books and moral leaders about social justice and morality is inconsistent with the fact and fiction regularly appearing on movie and television screens. The visual testimony is indeed more persuasive than the words it conflicts with.

Institutional Effects

Among the least visible of technology's collective effects are the impacts on institutions, social, economic, and political, which will probably prove to be more troublesome than environmental and other readily recognized consequences. As a case study of the aggregative effects technology can have on institutions, the next learning segment explores the challenges brought to government. The political system was selected for this purpose because the strains that technology imposes on it are particularly worth noting.

YOU DECIDE

In many segments of the learning program you have learned something about professional responsibilities of engineers. Tie all these comments together and add your own thoughts on the matter to produce a "statement on engineering professionalism." Don't overlook "after five" (off the job) responsibilities and opportunities.

Here is one person's characterization of engineering professionalism.

The truly professional engineer:

- Is, of course, technically competent
- Attempts to serve society's best interest by viewing problems broadly
- Keeps abreast of new developments that better equip him to perform his function as an engineer
- Remains alert to and concerned about the many ways his solutions directly and indirectly affect the public
- Balks at misapplication, poor taste, exploitation of and disregard for the public interest, and misrepresentation

- Implements his solutions so that social costs and disruption are minimized
- Follows up on each of his solutions to assure that it is successful and to learn from the experience
- Remains informed about key public issues, especially those related to science and technology
- Takes a stand; speaks out when situations in the community, the nation, and the world call for response from citizen-engineers
- Aids in the task of communicating technology, technological alternatives, and technological issues to laymen
- Participates in community affairs

SELF QUIZ

1. List the two levels of engineering's impact that were explored in this learning segment.

____

____

2. List four reasons why it is either difficult or impossible to include all second-order effects in the benefit-cost analysis of a proposed design project.

____

____

____

____

3. When it comes to human wants and needs, the engineer must have a certain competence and a certain attitude. What are these two things that an engineer must have in relation to wants and needs?

____

____

4. Briefly explain what is meant by the "impact of a specific technical venture" and how this impact differs from the "collective effects of technology."

____

____

____

____

____

1. Engineering has an impact at the level of individual creations and at the level of collective effects of all technology, or of a number of creations that converge in their effect on a significant portion of the population.

2. It is either difficult or impossible to include all second-order effects in the benefit-cost analysis of a proposed engineering design project, because (1) these "side-effects" are usually unintended effects, (2) these second-order effects are often not anticipated, (3) these effects are often the result of a combination of very many interrelated causes and effects that are virtually unpredictable, and (4) the victims of these effects are often not parties that are known or involved in the design process either as active instigators of the process or as beneficiaries.

3. The engineer must know the human wants and needs involved in an engineering project and must care about those wants and needs?

4. The impact of a specific technical venture is generally limited to a certain locality and definable segment of the population. The impact of a specific technical venture, even though it may well have second-order effects, is usually definable in terms of a limited, or at least finite, number of specific effects. The collective impact of technology, on the other hand, is not necessarily limited to any particular segment of the population, is not limited to any part of the globe, and has effects that may defy any explicit enumeration or definition.

Learning Segment 14
TECHNOLOGY, GOVERNMENT, AND DECISION-MAKING

LEARNING OBJECTIVES

- Understand that public officials must not only decide on overall priorities among large areas of technological need, but also have many options in each area from which to choose.

- Be able to explain the difference between type I and type II decisions.

- Be able to recognize and understand some of the handicaps that most governmental decision makers operate under when making decisions in technological areas.

TECHNOLOGY'S IMPACT ON GOVERNMENT

To appreciate the interaction between technology and government, you should be alert to a variety of relatively recent techno-political changes. Think for a moment about the total consequence of these developments: jet aircraft, computers, atomic bombs, television, rocket engines for space flight, intercontinental ballistic missiles, mechanical cotton pickers, and tape-controlled production equipment. Far reaching, to be sure, all blossoming during or since World War II. In aggregate these and numerous other post-1940 developments are significantly different, both in nature and in consequence, from pre-1940 products of engineering. If you view the long-term trends in the pace with which new capabilities are developed and applied, the power embodied in them, the scale of these endeavors, and other important qualities, you can't help concluding that there is a major discontinuity around the World War II period.

This change has meant an accelerated development and application of a more powerful technology in a world of increasing population density. One major upshot is that government and governing will never be the same. Modern technology has brought added opportunities, new strains, and different circumstances in a variety of other respects to all levels of government. You should know something about these effects, so here is a sampling:

There has been a major change in the types of decisions made by public officials. Now they must make many technical decisions involving the bulk of taxpayers' money. These decisions concern dams, highways, supersonic aircraft, ballistic missile systems, mass transportation, computers, water supplies, energy conversion, pollution abatement, traffic control, and weather satellites. No doubt you can continue the list yourself.

THINK IT THROUGH

Decisions must be made in order to allocate funds among such differing options as health delivery systems and tactical nuclear weapons. But even when it comes down to a particular engineering venture—a power generating and delivery system, for example—technology is so complex that the decision maker is still faced with difficult choices. Why is this so?

____

____

____

____

Some answers to this exercise appear in the following paragraphs.

For one thing, engineering continually enlarges the number of alternatives. Take the generation of electric power; within a few decades the number of conversion methods has tripled. In addition to several nuclear methods, we now know how to generate power by exotic-sounding processes such as magnetohydrodynamics, thermionic conversion, and thermoelectric generation.

Actually, when it comes to technical achievements, there isn't much that man can't do. The possibilities are almost unlimited. All we need is someone to pay the bill—and the price tags on many of these alternatives are staggering. The relevant question is no longer, "What can we do?" Now we ask "What *should* we do?" Obviously taxpayers are neither willing nor able to finance everything, so priorities must be established and choices made.

This leads to an enormously compex decision-making task for public administrators and legislative bodies, especially Congress, holder of the national purse strings. It must allocate a staggering sum of money to a limited number of undertakings, limited because, however large you may think that sum is, it can only support a fraction of the opportunities available. So public officials must choose. The matter is complicated further by the fact that a majority of the choices in one way or another involve science or engineering. Imagine yourself a member of Congress reviewing budget proposals submitted by various departments of the federal government. You are zeroing in on one narrow facet of the total budgetary picture, wrestling with questions such as: "How much can we justify spending on transportation?" "Of that amount, how much should be allocated to development of new capabilities, how much to application of existing capabilities?" "Of the money to be spent on developing new transportation modes, how much should go to space travel, how much to air transportation, how much to surface systems, and how much to subsurface modes?" (You may be picturing how complicated this becomes. Figure 1 will help you in this regard.) Of course, in competition with transportation for funds are all other technical opportunities, scientific research, welfare, foreign aid, education, maritime subsidies, U.S. Forest Service, and so on. The federal budget goes on and on.

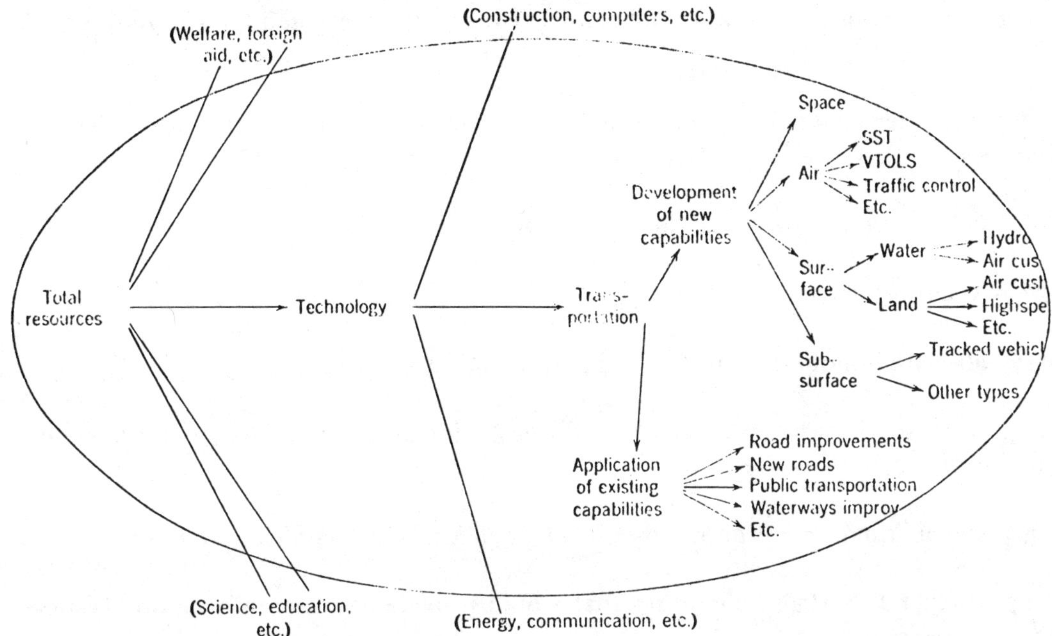

Figure 1. What resources should go to what alternatives? What opportunities should be foregone or postponed? Resources are limited. If this diagram were completed, it would require a space the size of a gym floor. It would, indeed, reflect the magnitude of the allocation problem confronting federal officials. The task is a staggering one partly because of the number of alternatives; everything seems to explode into numerous subalternatives, yielding thousands of choices.

THINK IT THROUGH

So far we have concentrated on decisions involving the federal government. State and local governments have their share of difficult decisions involving technology, too. Try to list a few.

____

____

____

Answers to this exercise appear in the following paragraph.

State and local officials will be quick to remind you that it is not only the federal government that is affected. You can imagine what it means to city officials to find industries emigrating and large numbers of unemployed persons immigrating. Of course, both trends owe much to machines of one kind or another. Furthermore, picture the headaches involved in:

• Deciding on the wisdom of a new connector between center city and the circumferential freeway.

• Locating such a roadway, especially when every alternative takes it through a densely populated area.

• Choosing between such a roadway and a public transportation system.

• Siting a new airport.

• Trying to alleviate interminable traffic and parking problems.

• Deciding what to do when the city's solid-waste incinerators no longer meet state-legislated pollution standards.

• Passing a new building code proposal that involves all kinds of technical standards.

• Deciding whether to install a computer that would provide a master information system for all city departments.

• Determining the economic wisdom of installing a system that converts sewage sludge into electricity.

The technical challenge to them is the combined result of technology-precipitated problems (such as pollution) and of new technological opportunities (such as computerized record keeping).

GOVERNMENT'S INFLUENCE OVER TECHNOLOGY

Modern technology affects government in many ways. But the reverse is also true; government has a big hand in determing the nature and role of technology. This influence is worth examining. Government cannot help influencing technology as long as it is by far the largest financial backer. The federal government buys weapons, trucks, computers, meteorological equipment, and so forth, the cost of which consumes a sizable chunk of the federal budget. In spite of the vast sums and therefore the influence involved, this is not the major cause for concern. More consequential is government's role in making the following types of decisions.

Type I Decisions

If man achieves a breakthrough in his ability to tunnel beneath the earth's surface, develops a practical means of generating electricity through nuclear fusion, or acquires almost any major new technical capability, it will be because he has chosen to do so. Today very few of these capabilities are simply "suddenly discovered." Their evolution is probably much less accidental and much more deliberate than you imagine. Ordinarily, a new capability is the result of a goal-oriented process that begins with a decision to commit financial and manpower resources to the development of that capability (for example, power from the atom). This is followed by a period of intense effort spanning months or probably years, usually resulting in the sought-after capability. These points are important enough to repeat: much of the new technology that evolves is planned, meaning that there is time to choke off technical ventures that will ultimately cause society more grief than benefit.

Consider the prospect of mass-producible, artificial, implantable hearts. It's an appealing thought: the extension of human lives by years through the replacement of defective hearts.

YOU DECIDE

Whether or not to develop a new capability to implant artificial hearts in a large number of people suffering from serious heart disease is a type I decision. How would you decide? Explain why.

____

____

____

____

____

Developing such a capability would certainly seem to further the humane, life-conserving objectives of medicine. But this may not be such a big favor to mankind at that. With it come some knotty social, legal, and moral headaches. Overpopulation is already an enormous problem. And if attempts were made to restrict the use of such devices to only the "needy" cases, where would you draw the line and who would rule in specific cases?

There is little doubt that such devices can be developed. True, engineers and medical researchers lack some of the necessary know-how at the moment, but this can be acquired. What's lacking is the funds and the decision to proceed. (What is your recommendation in this case: proceed or not?) The issue is whether or not to develop a new capability, referred to henceforth as a type I decision.

Type II Decisions

We have the ability to continue sending explorers to the moon, to dig a sea-level canal across Panama, to build a new freeway through any of our major cities, to provide artificial kidneys to all in need, to factory-build a sufficient number of moderately priced housing units. But should we do so? Such decisions, concerning the application of existing capabilities, are referred to here as type II decisions.

Here is one: It is now possible to maintain a central, computer-kept file that merges the previously separate files of social agencies, tax bureaus, police departments and other government agencies. A multiple-terminal time-sharing system makes this "data bank" possible. Should such a data bank be created?

LET'S BE SURE ABOUT THIS

This decision concerns the use of an existing capability; the computer equipment necessary for a national data bank has been around for awhile. Is this a type I or a type II decision?

Type II.

YOU DECIDE

A type II decision regarding whether or not to use our existing capability to merge the separate files of all governmental departments must be made. What are some of the advantages and disadvantages of an affirmative decision?

By eliminating expensive duplication of tiles, a national, state, or urban data bank can save taxpayers a lot of money. It can also improve the effectiveness of the sharing agencies; all law enforcement agencies, for example, can pool their information. In addition, such a comprehensive accumulation of data could be processed to provide invaluable statistics to social scientists and government planners. These and other benefits make the data bank idea attractive in some quarters.

But there is a risk involved: the possibility that such a system will be used to violate the privacy of citizens. It is conceivable that unauthorized persons or agencies could obtain confidential information. This and other potential abuses of citizens by this electronic "big brother" have stirred opposition among members of Congress, as well as among writers for the Sunday newspaper supplements.

WHO MAKES THESE DECISIONS?

Who is it that makes the type I decisions and thereby determines what technical possibilities become capabilities? Who makes the type II decisions and thereby determines how and where our technology is employed? What forces are at work in influencing these choices?

Government, business, and the public all have a say in these matters. Although the exact role of each is impossible to pinpoint, observers have no problem detecting major shifts in relative influence. They can tell you with confidence that:

Over the past few decades, government has assumed the major role as type I decision maker. It now makes most important decisions of this kind. The federal government now spends billions of dollars on the development of new engineering know-how that spans national defense, transportation, space exploration, atomic energy—you name it. Public funds allocated to research and development have increased dramatically since 1940. The prime significance of this deep involvement is that the federal government has become the prime determiner of the new technologies that man acquires.

THINK IT THROUGH

Can you think of any type I decisions that the federal government has been involved in over the years? Can you think of any technological capabilities that have developed as a result of the government deciding to allocate resources to the development of such capabilities?

____

____

____

____

Some answers to this exercise appear in the following paragraph.

Could workable, implantable artifical hearts become available without the federal government's backing? Possibly, but it is not likely. The VTOLS described earlier would be delayed years, probably decades, if their development depended on private financing and the demands of the consumer market. This shift in influence from private enterprise and the marketplace to government is significant, indeed. This is one of those major changes that schoolbooks have not caught up with. And how about you, are you still crediting private enterprise for significant developments that have evolved from government-financed R & D programs? You would probably be surprised to learn just how much of the digital computer's capabilities was paid for by the federal government. Time sharing, graphic displays, light pens, the core memory (on which computers depend so highly) and other hardware owe their existence to the "federal financial funnel." A similar story applies to jet aircraft.

Concerning type II decisions—government and big business now exert more influence on how and where technology is applied (as opposed to developed), the public less, than you would like to think. Most informed observers seem pessimistic about what effects the wishes of the citizenry have on where and how government uses its bulldozers, rockets, and computers. (In view of the complexity of these undertakings and of the issues involved this situation cannot be entirely otherwise.)

But, you say, surely consumers, by expressing their preferences at the marketplace, influence what big business does with its technical capabilities. Yet here, too, you may be naive. Today, mass persuasion techniques make grass roots consumer preferences almost a thing of the past. Consider the power of television in this respect. Do you believe that the auto industry has marketed 400-horsepower automobiles in response to demands by consumers? If so, you underestimate the behavior-influencing techniques of Madison Avenue. As a consequence, it is impossible to say what influence unadulterated consumer preferences have, because consumer behavior at the marketplace is so conditioned by the promotional process. Thus it appears that Everyman has less say than he would like—as citizen and as consumer—about the manner in which modern technology affects him.

While government assumes an expanding role in making type I and type II decisions, these highly consequential decisions fall to a disturbingly small number of people.

CHECK IT OUT

With the help of newspaper and newsweeklies over the past few weeks, identify examples of type I and type II decisions under debate in various governmental and corporate organizations.

Review your experience on the job. Can you identify any type I decisions your firm or organization has been involved in? A few type II decisions?

Answers to this exercise will depend on your research and experience.

HOW WE MAKE OUR DECISIONS

A substantial proportion of the federal budget is now allocated to scientific and technical ventures that are unknown or at best mystifying to the citizenry. As time passes, more and more multibillion dollar decisions are being made by a relative handful of people, while the electorate becomes more and more out of touch with the possibilities and considerations. The result is a new decision-making elite, consisting of persons in key government positions and their advisors in and out of government. These men have acquired unusual influence as a result of their access to technical and scientific knowledge that the masses no longer comprehend.

In the absence of public consensus–even awareness–this concentration of decision-making responsibility is hardly consistent with what the founders of this nation visualized. It is a serious matter that causes an increasing number of thinkers to wonder if our form of government is being slowly undermined because the electorate is unable to make intelligent evaluations in this age of complex technology.

CHECK IT OUT

For a revealing experience, take a major technical matter that Congress has had to cope with and research some (to do a comprehensive job could take months) of the debate, testimony, and issues associated with the decision. Possibilities: SST, ABM, Occupation Safety and Health Act, Environmental Policy Act, Alaskan pipeline, space shuttle, or automotive emission control.

____

____

____

____

____

Answers to this exercise will depend on the technical matter you choose to explore.

One of the first tangible signs of official awakening to inadequacies in this decision-making process was the appointment by Congress of a special fact-finding Commission on Technology, Automation, and Economic Progress. Some of this commission's concerns are apparent in these words from the first volume of its report, *Technology and the American Economy*. Published in February of 1966, many of this commission's concerns are valid today.

There are two questions before us. One is: What proportions of national income should go for what ends? The second is: How adequate are our mechanisms for making such decisions and assessing the consequences?

There is a widely held belief—derived from our experience in military and space technology—that few tasks are beyond our technological capability if we concentrate enough money and manpower upon them. However our relative affluence and technical sophistication should not lead us to believe that we attain all our goals at once. Their cost still exceeds our resources. Thus we will continue to face the need to set priorities and make choices.

We cannot set forth in this report what the specific priorities should be. We are concerned with *how* we decide what to choose. Congress has asked us: "How can human and community needs be met?" But there is a prior question: "How can they be more readily recognized and agreed upon?"

What concerns us is that we have no such ready means for agreement, that such decisions are often made piecemeal with no relation to each other, that vested interests are often able to obtain unjust shares, and that few mechanisms are available which allow us to see the range of alternatives and thus enable us to choose with a comprehension of the consequences of our choices.

Obviously the Commission felt, and many members of Congress and administrative agencies have so commented, that the absence of specific national goals and associated priorities is a serious handicap. Of course, we have goals, generally implicit, that have evolved over the years (security from aggression, for example). However, we have nothing resembling the deliberately established, comprehensive, operational set of goals that administrators and legislators should have. They deserve some sympathy; it is difficult to allocate resources intelligently as long as goals and priorities are so elusive. But this is not the only deficiency.

THINK IT THROUGH

Without clearly defined goals and priorities, it is difficult to allocate resources for the most rational options. But the fact that it is difficult to understand many of the complicated technical possibilities in either a type I or a type II decision can contribute to weaknesses in the decision-making process, too. How?

____

____

____

An answer to this exercise appears in the following paragraph.

Regrettably, in the allocation process, insufficient consideration is given to alternative courses of action. Too often, for instance, a decision is made to underwrite the development of a new capability without giving a fair hearing to other ways of accomplishing the same objective and other entirely different uses of the money. We have spent billions in public funds trying to develop an economical means of converting nuclear energy into electrical power. We have succeeded, but promising alternative sources of power have been known for some time, and they have received little serious consideration and negligible public funds for development. This is a familiar pattern. In many instances, this happens because one alternative has a champion—a federal agency, an industry, or a pork-barrel-conscious legislator—giving it an inside track over all others, regardless of merits and costs to society.

At present there is no satisfactory mechanism for assessing the total costs and benefits of alternatives and their relative contributions to national goals. Members of Congress have no adequate source of answers to questions like: "What would be the long-run costs of the SST?" "The benefits?" "The indirect effects?" "The risks?" "How does it compare with alternatives?" The SST was chosen to illustrate this point because, during the 1971 Congressional deliberations that resulted in termination of the United States SST project, we saw a scandalous display of ignorance, bias, distortion, lobbying, emotion, unsubstantiated claims, media-fanned scare stories, conflicting testimony, and just about everything else you would not want in or surrounding such deliberations. This is what happens when the stakes are so high, the issues so burning, and the facts so sparse. What a way to decide!

THINK IT THROUGH

In the learning segment you have been confronted with a sample type I question about artifical hearts and a typical type II question about a national data bank. Did you feel you were equipped to answer these questions? What are some of the things that one must have in order to cope with them?

Did you feel that you were equipped to answer the type I and the type II questions asked above? What are some of the things that one must have in order to cope with such questions as the artificial heart and the national data bank?

____

____

____

____

____

____

Some answers to this exercise appear in the following paragraph.

You are not equipped to answer those questions until you know, specifically, the benefits expected from each venture, total costs anticipated over the long, run, their indirect effects, and alternative uses of the resources required. In fact, before a commitment is made, you should also know what kind of difficulties to expect in controlling the use of artificial hearts and abuse of a national data bank. Similarly, unless public officials and business executives have this same kind of information, they too are ill-equipped to make such decisions. But they do it all the time. Certainly decision makers are entitled to some mistakes, even an occasional blunder; these are very complex decisions. But we do have a right to expect them to ask the right questions, to obtain better information before making technical decisions affecting millions of people and requiring major commitments of resources.

The People Involved

How qualified are members of Congress and other public officials to make such decisions? In Congress, for instance, the number of legislators with a technical education has been running at about a half dozen. In other branches the proportion is not much different. So you can hardly say public officials, as a group, have a background commensurate with technically demanding decisions. By their own admission, they are ill-equipped.

So, you say, if most officials are not qualified, and if comprehensive, objective information is not really available, why not call in "experts" and rely on their advice?

THINK IT THROUGH

Does Congress call in experts and rely on their advice? What is the real problem here?

____

____

Answers to this exercise appears in the following paragraphs.

Congress does rely heavily on expert testimony, but this reliance turns out to be a prime source of disenchantment with the whole decision-making process. Most experts who come before Congress to offer testimony on costs, benefits, and risks associated with some technical or scientific venture are biased. Representatives of federal agencies usually provide overly optimistic appraisals of their proposed programs at Congressional budgetary hearings.

Well, you say, why not rely on experts outside government to provide Congress with appraisals of technical ventures proposed by administrative agencies? Congress does this, but these advisers have their biases, too; they are discipline-oriented if not agency-oriented. Biologists will testify in favor of almost anything that will improve the environment; authorities on urban affairs feel that not enough money can go to the cities; aerospace people are appalled that we are not spending more on space exploration; and so it goes.

Here is an illustration: Not long ago a proposal to drill through the Earth's crust to the mantle, known as the Mohole Project, was the object of much discussion in Congressional hearings. Over the course of prolonged testimony, there were references to Russian efforts to do likewise, creating the impression that the two nations were involved in a "race to the mantle." Yet there was no evidence to confirm this and, since then, observers have concluded that the supporters of this project let their enthusiasm get the best of them while they were "advising" Congress.

The elaborate technical advisory system now extensively employed by the Executive and Legislative branches of the federal government is vulnerable to criticism, and has gotten it. The upshot of these agency, mission, and discipline biases, after reams of testimony preparatory to a technical decision, is that officials must do a lot of second-guessing.

Of course, inviting a progression of experts to testify on the merits of a proposed engineering venture opens the door to spokesmen for special interests.

Unsolicited "Advice"; Lobbies; Industrial Constituencies

Vested interest groups, lobbying, and pressure politics are not new to you. But you may be unaware of the amount of pressure that government agencies themselves exert on legislators to sway them in favor of pet projects. The power plays, the politicking, and the infighting are elaborate and extensive.

Each federal agency has a powerful source of pressure in the research institutions and manufacturing organizations that derive large financial benefits from the agency's activities and programs. This is a source of dismay to many citizens. Nowadays, an agency simply hints that one of its proposed programs may not get the financial support requested from Congress, and its followers will set their lobbying machinery in motion.

Much has been written in the recent past about the followings that technical superagencies have cultivated. The most widely quoted commentary was made by President Eisenhower. He was warning the nation—and he was in a position to know—of the existence and dangers of a military-industrial complex which, in his words, is a "conjunction of an immense military establishment and a large arms industry," exerting an influence "felt in every city, every state house, and every office of the federal government." He called on the nation "to guard against the acquisition of unwarranted influence."

The primary significance of these powerful constituencies of federal agencies is that they have much at stake in certain proposed engineering projects. When such a proposal is under discussion in Congress, they throw their weight behind the sponsoring agency. When multibillion dollar contracts are at stake, you can be sure there is some sweet and fast talk in the smoke-filled back rooms. Unfortunately, there is no counterbalancing pressure representing the best interests of society.

LET'S BE SURE ABOUT THIS

Our representatives and senators in Congress work under many handicaps in trying to make the decisions that allocate a great share of this country's resources in the federal budget. We have discussed many of these; list three.

___

___

___

Some answers to this exercise appear in the following paragraph.

Some of the handicaps members of Congress and other allocators of public funds must work under are: their own nontechnical backgrounds; inadequate and often misleading advice; unsatisfactory benefit-cost appraisals; and pressures from vested-interest groups. The result is that we have been led down some unprofitable, sometimes hazardous, technological paths. But let's not be too hard on the legislators; the supporting systems (for example, the information resources and appraisal mechanisms) are more at fault than the individuals. That system is geared more for the relatively simple, slow-moving, sparsely populated world of two centuries ago, when man's power to control nature and other people was insignificant compared to now.

So much for the bad new—the limitations of the means with which this nation makes its major techno-political decisions. There is cause for concern. Yet there is some partially compensating good news; Learning Segment 15 explores it.

FIGURE IT OUT

You are a staff assistant to the administration of a large city. You have been assigned to investigate major mass transit alternatives and to identify criteria (officials would probably say *factors*, but you recognize that functionally these are *criteria*). Officials want you to identify the modes they should consider and to provide a comprehensive list of criteria that they should consider in selecting the final system. List some alternatives and criteria in the space below. If you do your job well, you will have an impressive array of alternative and a long list of criteria, which together indicate the complexity of the decision facing officials.

Alternatives: Conventional buses, flywheel-propelled buses, electric buses, steel-wheel trains, rubber-tired trains, suspended trains, air-supported vehicles, and so on.

Criteria: Right-of-way, equipment, and construction costs; operating costs; reliability, safety, carrying capacity, travel time, comfort, noise, visual impact on the environment, effects on air quality, effects on land use and land values, effects on peripheral traffic arteries, benefits for nonusers of the system, flexibility, type and amount of energy consumed, and so on.

CHECK IT OUT

Try this same approach to some problem you or your boss might be facing on the job. If it turns out well, use it to help your organization find a suitable approach to the problem.

SELF QUIZ

1. Government officials must decide on priorities in the allocation of funds for technological solutions in such areas as public works, health delivery systems, and energy conversion. But even when these overall priorities are decided upon, officials are faced with difficult decisions regarding any particular area, for example, power generation. Why is this?

2. List three technological areas that might present difficult decisions to state and local officials.

3. Explain the difference between type I decisions and type II decisions.

4. List four handicaps that our representatives and senators operate under when faced with complex decisions involving technological matters.

1. There is difficulty not only when assigning overall priorities to broad technological areas of need, but also when these priorities are decided. This is because there remain a great many options in each technological area, mainly because of the constantly accelerating pace of technological activity and the creation of new designs. For example, within a few decades, the number of methods for converting basic forms of energy into useful electricity has tripled. In addition, decision makers are hampered by grossly inadequate evaluations of the alternatives available.

2. Technological areas that might present difficult decisions for state and local officials would include traffic planning and highway design, public transportation design, airport or seaport design, solid waste disposal, sewage disposal, building codes, computerized information systems, and so on.

3. Type I decisions concern whether a new capability is to be developed in some technological problem area. Type II decisions concern whether and where to apply existing technological capabilities.

4. Some of the handicaps that representatives and senators in Congress must operate under are: (1) their own lack of technical expertise, (2) inadequate advice, (3) misleading advice, (4) the inability of anyone to provide fully adequate benefit-cost analyses of most major technological proposals, (5) insufficient time and resources to give full consideration to all available alternatives, and (6) pressure from a host of lobbying and other special interest groups.

Learning Segment 15
THE ASSESSMENT AND MANAGEMENT OF TECHNOLOGY

LEARNING OBJECTIVES

- Be able to explain the basic elements of technology assessment.
- Be aware of recent increases in capability and effort in the area of technology assessment.
- Apply the basic steps of technology assessment to typical technological decision situations.
- Recognize some of the types of restraint that might be used to manage technological development.
- Understand the need for proper timing in the enactment of legislation restraining the application of new technology.
- Be able to weigh some of the trade-offs in legislating restraints on the application of technology.
- Recognize ways in which engineering and related activities are subject to legislative restraints.

TECHNOLOGY ASSESSMENT

What might be done to improve government's means of making type I and type II decisions? You could make some appropriate suggestions simply on the basis of reason and the contents of the preceding learning segment.

YOU DECIDE

What suggestions would you make to improve our government's means of making type I and type II decisions?

____

____

____

Some answers to this exercise appear in the following paragraph.

1. Develop a strengthened mechanism for establishing national goals and associated priorities. Do you agree—is it difficult to make means choices when ends are so vague?

2. Enable public decision makers to consider more alternatives before committing resources to a particular one.

3. Involve more persons with technical or scientific backgrounds in government, particularly as elected officials.

4. Develop a mechanism that will provide officials with objective assessments of the full impact of proposed technical ventures (in other words, with information on such matters as costs, benefits, and risks—information that they have been lacking).

There is no significant progress to report for the first three of these proposals, but for the last there is. The good news is the recent evolution of an evaluation process referred to as *technology assessment.*

Many authorities are using this expression to refer to an impact analysis of a technical undertaking. Although there is general agreement that technology assessment will serve man well, in this early stage, when it is still taking shape, there is no such agreement on what a technology assessment should consist of. So the following characterization of technology assessment, presented in a hypothetical illustration, is a composite based on the dialogues now taking place in the journals and meeting halls.

Areas for Questioning

A jetport has been proposed for an offshore site not far from a large metropolitan area. This facility will be pile-supported, floating, land fill, or polder; the details have not been resolved. Before proceeding with this venture, a full appraisal should be made of its social, economic, and environmental effects. Technology assessment involves all of that. It calls for careful consideration of:

1. The *appropriateness* of additional air transportation facilities. Is there a *need* for another airport in the area in view of such considerations as the projected numbers of travelers and the possibility of new modes of transportation that would significantly affect the volume of jet traffic? There is a related matter, one that is rarely considered and is obviously more difficult to cope with. It has to do with the *wants* (desires if you prefer) of society. Strictly

speaking, we don't *need* more jetports; we build them because a sufficient number of people presumably value speed and convenience more than the things (like a cleaner, quieter environment) that must be sacrificed to obtain those benefits. This is a trade-off that should be reevaluated each time expansion of the air travel system is considered, since it probably will not hold up indefinitely. To put it another way, the air travel interests will tell us that we *must* continue expanding air terminal facilities to accommodate future passenger volumes. Although it is rarely done now, eventually more people will challenge the assumption underlying this "expanding indefinitely policy." Those who do are finally viewing what are usually cited as "needs" for what they really are: wants. These are complicated matters, true, but that hardly seems sufficient reason for neglecting them while type I and II decisions are made.

2. The *alternatives*. If expansion of the city's air terminal facilities is appropriate, it would be foolish not to consider other means along with the offshore proposal.

3. The *benefits*. In this example the benefits will be safer operations and less noise at the city's existing airport, assurance of adequate air travel facilities for some time to come, stimulus of the local economy, and so forth.

4. The *costs*. The obvious costs components are construction and operation. However, other less tangible costs are likely, many of which are environmental and which, in turn, reduce other use values of the region.

5. *Other effects*. These cannot be identified in advance as positive, negative, or even as consequential. This facility will affect navigation and land use; it will probably affect land values; it could affect air and water currents. The magnitudes and the consequences of these effects can be predicted only after in-depth investigation—if then.

6. *Restraints necessary*. If restrictions on flight paths are forthcoming, if noise limits will eventually be imposed, if a minimum is desirable for the distance between jetport and shore, these should be established before the decision to build. These restraints most likely will affect the feasibility of the project.

YOU ASSESS IT

On the basis of current usage and the number of ships that are too large for the current Panama Canal, a new and enlarged canal is the same general area seems inevitable. This project is mainly in the talking stages at present, and most of the talk is about a sea-level canal. The present facility has locks and a freshwater lake that offer a barrier to the flow of sea life, but the sea-level version will permit the free exchange of species between the Atlantic and Pacific Oceans. This is one of a number of significant effects of this proposal.

This venture is a natural for technology assessment. Draw up the specifications of such an assessment, indicating what questions you believe should be answered before a decision is made. Assume that you are instructing a team that is about to undertake an assessment of this venture. Refer to the January 29, 1971, and September 26, 1969, issues of *Science* for some helpful information.

Appropriateness

____

____

____

Alternatives

____

____

____

Benefits

____

____

____

Costs

____

____

____

Other effects

____

____

____

Restraints

____

____

____

Appropriateness: What volume of ship traffic is predicted for many decades ahead? Will the vessels of the future call for canals as we know them? What is the penalty if the present canal is not replaced or enlarged?

Alternatives: Why not enlarge the present canal or build a second one parallel and close to it? Can vessels be towed faster to increase capacity? Is nuclear excavation feasible?

Benefits: What are they? Who benefits?

Costs: Land and construction? Operations?

Other effects: Environmental? Political? Military? Biological?

Restraints: What about limits on the size of vessel the new canal will accommodate?

Need to Coordinate Multiple Technological Events

For introductory purposes certain features have been omitted from these hypothetical examples that surely should be included in real assessments. For instance, the airport case above sounds like a decision in a vacuum. Yet technology assessments must transcend the specific ventures themselves, because so many of these decisions are interrelated. Many technical undertakings affect the same atmosphere; many consume a given resource like oil; many can affect employment levels in a given area or occupation.

THINK IT THROUGH

What would be wrong with governmental officials simply considering each technological proposal independently of all other technological activities either being proposed or in progress?

____

____

____

____

____

An answer to this exercise appears in the following paragraph.

If decisions concerning individual projects are made independently, government will be continuing its practice of creating new problems with its solutions to old ones. The interrelatedness of techno-political problems and decisions continues to increase. Surely you have observed how the so-called energy problem interacts with problems of economic growth, transportation, solid-waste management, environmental quality, balance of payments, and no doubt others. Yet interrelatedness is a characteristic of problems men are poorly equipped and seldom inclined to cope with. This surely is a prime challenge to technology assessors.

Need to Consider Multiple Effected Populations

Let us hope that technology assessors will come to grips with the "conflicting constituency" problem. It's a headache to cope with and, consequently, tempting to slight. Comparing benefits with costs in the tire-mounter case is a relatively simple matter; comparing them when one segment of the population derives most or all of the benefits while another bears most or all of the costs is a much more difficult problem. Consider the bank that is installing a computer system. A primary benefit is improved service to bank customers; a major cost is the personal hardship suffered by bank employees displaced by the computer.

CHECK IT OUT

Consider your own experience on the job or in your community, and list several examples of technological projects or proposals that affect different parts of the population differently.

____

____

____

Answers to this exercise depend, of course, on your own experience.

There are many such multiple populations problems. The same kind of situation prevails in siting decisions for airports, highways, and power plants. Often there are more than two population segments that are affected differently and, to complicate the decision further, there is usually some overlap among these "conflicting constituencies." There is no easy out for decision makers on this score; this is another prime challenge to technology assessors. What we *can* expect, however, is that such constituency conflicts will be explicitly and realistically recognized in the assessment process.

To be of maximum benefit, technology assessments must incorporate a "global" view; these decisions often produce consequences spanning many political boundaries. This aspect of assessments is likely to be slow in developing. After all, towns are still willing to discharge sewage for towns downstream to worry about. So what can we expect from nations?

Need to Look Forward to the Future

Technology assessment need not—hopefully, it will not—be restricted to a reactionary role. Surely there are highly desirable capabilities that society has not benefited from because the economic incentives or the necessary institutional arrangements are lacking. Technology assessments rightfully can be expected to alert government to such opportunities.

Therefore, technology assessment is intended to supply much of the information that public officials need to make intelligent technical decisions. Although different authorities have somewhat different view of the character of technology assessment, most agree that, at a minimum, it should predict direct and indirect effects over the long run from the point of view of society at large. Note particularly the future orientation of technology assessments; it is primarily a means of predicting wants, costs, and benefits; of anticipating the need for regulatory legislation; of sounding early warnings of impending hazards. This future orientation and the comprehensiveness of technology assessment seem to be its prime virtues.

LET'S BE SURE ABOUT THIS

Technology assessment is not intended to be a reactionary resistance to technological innovation. We have just stressed three ways that it is future oriented. List them.

____

____

____

Technology assessment is a means of (1) predicting wants, needs, costs, and so on; (2) anticipating the need for restricting legislation; (3) sounding early warnings of impending hazards.

Developing Capability in Technology Assessment

Although the term and its comprehensiveness are new, many of the specific concepts and practices that comprise a technology assessment are not.

A close relative and recent forerunner of technology assessment is the provision of the 1969 National Environmental Policy Act requiring environmental impact statements from all federal agencies. That act, incidentally, requires federal agencies to file a statement with the Environmental Protection Agency whenever a technical venture is contemplated. Thus, before the Corps of Engineers can start construction of a new dam, or before a proposed highway can be approved for federal funding by the Department of Transportation, a comprehensive statement must be submitted, covering environmental effects, irreversible commitments of resources, and alternatives to the proposed project. Observe that what the Environmental Policy Act requires for environmental impact, technology assessment provides for total impact.

Although the term and the concept of technology assessment originated in the late 1960s, by the early 1970s it was already widely known and discussed in political, scientific, and technical circles. The speed and extent to which the notion has spread indicates the widespread concern for the problem it is a response to, and the hope that many observers see in this approach. In fact, Congress recently voted itself a modest technology assessment capability in the form of an adjunct Office of Technology Assessment. A significant and welcome development indeed. Now members of Congress have an agency that has no specific technology to promote, staffed by professionals whose business it is to make independent appraisals of proposed technical ventures, to which they can turn for the information they need.

The appropriateness of Congress's action in creating it's Office of Technology Assessment becomes more apparent as you learn more about the relative roles and resources of the Executive and Legislative branches in the techno-political decision-making process. Projects like the supersonic transport, the breeder reactor, and military weapons systems originate with agencies of the Executive branch and are sent to Congress for authorization of funding. Congress rarely receives a satisfactory explanation of the alternative programs considered and why these were rejected in favor of the proposal ultimately selected. Nor are such programs accompanied by background papers stating the important technical considerations, evaluating the benefits and costs, and identifying the risks arising from lack of technical information or understanding. When such studies are performed by Executive agencies, they are rarely made available to Congress (let alone the public). So up until now, not having the time or staff to perform parallel investigations, Congress has not been able to fulfill its "checks and balances" function. The Office of Technology Assessment, if adequately financed, will enable Congress to respond more intelligently.

Interest in technology assessment is not restricted to the federal level. There is talk among state and city officials about it, and surely some of them will acquire this kind of capability.

LET'S BE SURE ABOUT THIS

List two recent developments related to developing capability in the area of technology assessment.

____

____

Two recent developments contributing to an overall capability in technology assessment are: (1) the enactment of the National Environmental Policy Act of 1969 and (2) the establishment of Congress's adjunct Office of Technology Assessment.

The Prospects

Certainly the need exists and surely technology assessment is potentially a very useful instrument. Whether it evolves as such is another matter; it is still in the formative stages, and it has its skeptics, even opponents. There are those who argue that the present system has survived the test of almost two centuries of use; why tamper with it? And, too, there is some fear that rationalization and systemization are taking over from good old-fashioned politics. But when multibillion-dollar investments are involved, the physical well-being of millions is at stake, and far-reaching and of times irreversible social and environmental effects are inevitable, do you want lobbies, and pressure politics to have the influence they now have? Other opponents see technology assessment looming as a pinch in the pipeline of progress and, for that reason, they are prone to refer to it as "technology arrestment."

Technology assessment is no panacea. Although it will provide significantly better information than is now generally available to decision makers, surely it is not going to eliminate mistakes. Appraisals of costs, benefits, and risks are not going to be completely factual. Furthermore, there is no need to fear the elimination of human inputs—there will still be call for judgment, debate, compromise, and expert opinion.

Up to this point government has received most of the attention, which may well prompt you to conclude that private enterprise is getting off too easily. But it is not off the hook yet. Furthermore, there is sufficient reason for our emphasis on political decisions in the technical area; government has become more influential than private enterprise in many such matters. A majority of the most significant and controversial engineering ventures today—including new freeways, urban mass transit, proposed jetports, weapons systems, SSTs, urban renewal, and space exploration—are government doings.

MANAGING TECHNOLOGY

Choosing what capabilities to develop (type I decisions), deciding how to apply existing capabilities (type II), and determining what we dare not do with them (type III) are the main decisions through which man manages his technology. The term *technology management* and what it implies are indeed appropriate. Hopefully, both will soon be widely accepted. In addition to the three decision types, it includes feedback mechanisms that enable decision makers to alter their policies, regulatory practices, and investments in research and development, as the economic, social, and environmental well-being of society dictates (Figure 2).

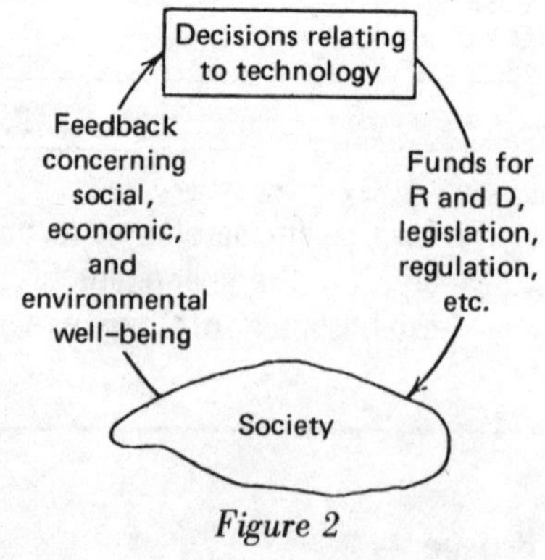

Figure 2

What governmental structures will best perform this management function is a complicated matter; thinkers will labor over this matter for some time. However, it is clear that careful planning, effective feedback, a more distant time horizon than decision makers are accustomed to using, and a constituency of men everywhere are essential features of a successful technology management capability.

Responsibility in the Private Sector

In the private sector the counterparts of public officials who make type I and type II decisions come in for their share of criticism. But for some different reasons. Unlike their political counterparts, decision makers in private enterprise can neither readily shy away from the matters of goals and priorities, nor be consistently guilty of tunnel vision with respect to alternatives. Neglect these matters for long and the management, if not the whole business, will go.

The biggest failure of management is in cost accounting, using the term in a broad sense. Manufacturers ordinarily let society assume the cost of properly disposing of the wastes they dump into streams, lakes, and atmosphere. Who do you think strip-mining companies have assumed will bear the costs of devastated countryside? Similarly, the cost or properly disposing or recycling of consumer goods at the end of their useful life is ordinarily charged to society. This practice of ignoring certain "external" costs for which they are never billed affects producer and consumer behavior. Many of their decisions would be affected if buyers paid true costs. The true cost of electricity would include what you are now paying, plus an assessment for all environmental damage inflicted from the coal mine to the outlet in the wall, plus any other costs now being borne by someone other than the consumer. Many people would behave differently if they simply knew the true costs of items like electricity or their automobile. There is no danger of that, however, for all practical purposes true costs are unknowable.

Costing practices are not the only criticizable aspect of type I and type II decision-making procedures in the private sector. There are businesses that market some pretty lethal products, and surely there are other ways in which private enterprise abuses man and nature through technology. But most of these abuses can be traced back to their particularly vulnerable costing practices.

There may be some who feel that private enterprise can eventually be persuaded to assume the costs of the public resources they consume, pay for the damage they do, and otherwise absorb costs that they have heretofore "externalized." The more realistic critics of businesses that palm off sizable costs on society think otherwise.

The Need for Restraint

Man has acquired some remarkably powerful capabilities that often prove to be double-edged. The bulldozer has certainly made its mark as a benefactor *and* devastator, and television, which is a great source of entertainment and enlightenment, has frightening potential for influencing public opinion and social values. The device for releasing the power of the atom is perhaps the most dramatic of man's many double-edged contrivances. Although not inevitably so, these and other technical developments do, in fact, spell ill for at least some of mankind. Even contrivances born of the best of intentions sometimes are mixed blessings.

So technology, while it provides man with an opportunity to better his life and lot, has increased the variety and potency of his opportunities to abuse. To abuse in several ways, in fact. One is the infliction of harm directly and deliberately, as with weapons. The majority of technological abuses, however, are of man and nature as by-products of the use of technology for otherwise innocuous purposes (for example, smoke from the factory's stack).

Commercial deep-sea fishing has become so efficient, thanks to technology, that relatively few fishermen are now able to sweep the oceans free of commercial fish. And this is what the fishermen of all nationalities seem bent on doing. Engineering has provided them with tools that serve an important and highly useful purpose. However, they can also use these tools to destroy a valuable resource. Such exploitation is hardly considerate of future fishermen, to say nothing of future, underfed masses. What fishermen are doing to fish resources, you can see other private interests doing to public resources like the atmosphere, waterways, and land itself. This is nothing new to you.

A major upshot of man's enlarged capacity to commit "technological sin" is the need for greater restraint in what individuals and institutions do with their machines and structures. The streets are more crowded, and so are the highways, lakes, rivers, and airways. With planes, guns, automobiles, industrial effluents, and so forth, it is the increased power to inflict harm combined with greater population density that create the need for more restraint.

True, too, we are much more dependent on technology for fulfillment of needs and wants; the public is now at the mercy of a variety of technically based enterprises, power companies for example, so some government regulation of them is necessary. Indeed, this has already come to pass for some industries.

But power, population, and dependency are not the whole story. There is this matter of complexity; it continues increasing, conspicuously so in consumer products; the automobile, washing machine, and even the bicycle are illustrations of this. As a consequence, it is becoming more and more difficult for the buyer to appraise intelligently what he is about to purchase, to anticipate hazards in operation, and to make repairs and adjustments. Thus, the consumer is more often at the mercy of the producer.

And so technology provides more than ample opportunity to abuse; in fact, when it comes to powers to contaminate, maim, clutter, and influence, man has never had it "so good." Who disagrees that some restraint is necessary? The issues are how much and what kind. Concerning how much: there is a sound case, based on the power, population, dependency, and complexity arguments, for greater restraint. Certainly there are strong political pressures in that direction. But how far we should go in this respect is going to be a rough problem to resolve. This question, and questions concerning the most effective type of restraint (self versus legal versus any other restraining mechanism that shows promise), we refer to as type III decisions.

TECHNICAL TERM

Write a brief definition for the following term.

Type III decision

___

___

___

Type III decisions concern such matters as how much and what kind of restraints should be applied to technology, as well as what kind of restraints are most effective and proper in a given situation.

Mounting Pressure

A conspicuous manifestation of the prevailing feeling that man must exercise greater restraint in what he does with his technology and where he does it is the growing resistance that citizens are offering to certain technical ventures. Certainly those who misuse technology are hearing about it, especially from the environmentalists and consumer advocates. The resisters are becoming more numerous, more vocal, better organized, and more effective.

These voices of protest serve a restraining function. There are proposed power plants, pipelines, freeways, and other projects that are being resisted by ad hoc citizen groups. In aggregate, these opposition groups serve as a retardant, but not necessarily the best kind. Fighting virtually every proposed construction project on an ad hoc basis has been a useful interim means of restraint. However, among the drawbacks of this approach are the prolonged legal proceedings associated with these bouts between backers and opponents; they are trying and expensive to the combatants and society. This is one reason many are unhappy with this method of restraint.

THINK IT THROUGH

Can you think of any other reasons why ad hoc opposition to particular projects may not be an effective and productive restraint on technology?

___

___

___

Some answers to this exercise appear in the following paragraph.

There are other disadvantages to the ad hoc opposition to major, one-shot technological acts. The approach is a restraint mechanism but is unsuccessful in some instances, inapplicable in others. Citizen opposition by itself has not successfully curbed pollution by industry, municipality, or automobile. There was little response to protest until legal restraints were enacted and offenders threatened or prosecuted. Furthermore, there are many misuses of technology that are repeated acts by individuals with their automobiles, snowmobiles, and guns, to which direct citizen opposition, practically speaking, does not apply.

Alternative Restraint Mechanisms

There are a number of ways of preventing something like a complete wipeout of the sea's fish resources. One is self-restraint–adherence to sensible conservation practices by commercial fishermen. An alternative is the "bill them" approach, under which fishermen are taxed according to their catch and these funds used to support fisheries and other means of sustaining fish populations. Another is enforcement of legal quotas, intended to conserve this resource, by an international agency. These three alternatives–voluntary restraint, billing abusers for damage done, and government-imposed restraints–present themselves in most instances where technological abuse is likely.

YOU DECIDE

Which of these three alternative restraint mechanisms–voluntary restraint, billing for damage, and government-imposed restraints–do you think will predominate in the long run? Explain why.

____

____

____

____

____

A brief analysis of several opinions appears in the following paragraphs.

You probably have an opinion, and so do many others. Here is where the weight of recorded opinion lies: Virtually everyone is pessimistic about self-restraint. Most agree that the opportunities for abuse provided by technology are too numerous and too tempting; too much so for humans, frail as we are. You cannot depend on individuals to refrain voluntarily from using these power to do their fellow men in or to do them out of something. And it is naive to expect business interests to refrain voluntarily from using technology to exploit national resources or the public. Given the opportunity, man will exploit man and man will exploit nature. Of course, that opportunity has always existed, but never were the means for exploitation what they are today.

"Billing them" is a possibility. Through taxation or a more direct means of assessment, abusers pay for damage done. An example is government assuming the responsibility for restoring land defaced by mining companies and billing the companies accordingly. This is not a fine; here the abuser may continue defacing the countryside or polluting or whatever, but he pays. Although this method has a certain appeal (it is a step toward true costs), it receives little support in the literature or in practice, probably because it is a cumbersome mechanism and because of the difficulty of assessing the damage (for example, to residents near a jetport).

So if there is so little faith in voluntary restraint, if economic disincentives are considered impractical, what prevails is the method of last resort—legislated restraints.

Legal Restraints

Governmental restraints on the use of engineering contrivances are nothing new, nor are regulatory agencies for this purpose. Speed laws have been around for some time; regulation is certainly not new to the electric power industry; there already are laws controlling the use of electronic bugging devices. The issue is one of degree, touchy to be sure, that will not be resolved overnight or without controversy.

The pressure on government to legislate against technological sins is mounting. The cry for tighter pollution controls has been extended to auto features, bugging devices, motorboats, guns, and a wide variety of other consumer products. Tighter curbs have also been demanded on private exploitation of scarce resources like water, the marshes, minerals, fish and fowl, and open land, most of which are consumed by engineering creations of one type or another. Tighter safety regulations in mines, factories, and construction, where machines are used extensively, are also being requested. Tighter legal restraints appear to be the only practical alternative in many instances. With this prospect, it is important to say something about two key decisions legislators face. One concerns timing, the other degree.

Setting legal restraints—the timing. The snowmobile provides a timely illustration of what is becoming a familiar pattern of events. While this "sport" is becoming popular, participants are relatively free to use their machines where and how they please, at least long enough to become accustomed to this freedom. But not without unfortunate consequences for man and nature, so that some restrictions on features and uses of snowmobiles seem inevitable. You can imagine, though, how these would be welcomed now.

What is happening with the snowmobile is getting to be a familiar story. It begins with the acquisition of a new capability (for example, for transporting gas through a pipeline or mechanically harvesting fish), the use of which is practically unrestricted, at least at the outset. But exploitation creeps in and continues unabated, perhaps unchallenged, until society eventually and belatedly responds by pressing for government controls to curb abuses. But this time lag between the time a capability becomes available and the imposition of restraints causes big trouble. Typically and understandably, the individuals or institutions using that capability have become accustomed to freedom in doing so; a precedent has been set.

THINK IT THROUGH

The arguments applied by pipeline companies, mining operators, or automobile makers, to include just a few examples, will go beyond the mere citing of precedents. What do you suppose is one of the most important considerations in their cases?

____

____

____

An answer to this exercise appears in the following paragraph.

In the case of pipeline companies, mine operators, or automobile makers, it is more than a matter of precedent. After it has invested millions, an industry is not eager to rebuild the system according to a new set of rules. This is true of the airline industry; although their engines are noisy, it would be unreasonable to now expect them to invest hundreds of millions to replace them with quieter engines, even if they were available.

The point is that we should be doing a more thorough job of studying new capabilities as they evolve, in an attempt to anticipate the eventual need for restraints, so that:

- The cost of enforcement can be predicted and included as part of the total cost of a new capability, thereby influencing the decision to proceed with its development. Hopefully we will not invest in development of a technical capability if we know in advance that it is going to be miserable or impossible to live with.

- The necessary regulatory legislation can be enacted before large capital outlays are committed, before "rights and privileges" become established, and before serious damage is done.

Thus, there is another important purpose to be served by technology assessment: anticipation of the need for restraints, for costing purposes and for the establishment of laws that will be on the books when the new capability blossoms into widespread use.

LET'S BE SURE ABOUT THIS

In what two ways, just discussed, does technology assessment promise help in setting the timing of legislation restraining the use of technological capabilities?

____

____

____

____

____

Technology assessment can help in predicting the total cost of a new capability, thereby influencing the decisions to go ahead and enabling laws to be passed that guarantee that the true costs are borne by those who benefit from the technology.

Technology assessment can also give advance warning of adverse effects to be expected from application of new technology, so that restraining legislation can be enacted before anyone engages in precedent-setting exploitation of the technological capability.

Setting legal restraints—the trade-off. Sooner or later, like it or not, society will be forced to accept closer regulation of such matters as what private and public enterprises discharge into the environment, what materials manufacturers can use, and of how much energy we may consume. But there should be a limit; underrestraint can eventually turn to overrestraint. In some areas this appears to have occurred already.

YOU DECIDE

One of the more curious and controversial restraints on man's use of his technology is the popular requirement among states that motorcyclists wear helmets. Can you explain it? Can you justify it? What issues does this requirement raise?

____

____

____

____

____

Some answers to this exercise appear in the following paragraph.

The motorcycle helmet law prevalent among states is generally viewed as overrestraint. And what about the seat belt interlock fiasco in 1975 automobiles? Or bring up the matter of governmental regulation with an executive of a railroad, electric utility, or airline. Obviously a trade-off is called for, a balance between ridiculous extremes, one being a laissez-faire policy that permits us to hang ourselves, and the other being an overrestrictive policy that stifles human freedom and economic initiative. Specifically, the trade-off is: the cost of physical, social, psychological, and other damage we do with our technology versus the cost of creating, enforcing, and complying with restraints, as well as of loss of freedom.

APPLY YOUR KNOWLEDGE

In going through an optimization procedure, trading off various criteria involved, what is the main objective in optimizing the amount of legislated restraint?

____

____

____

An answer to this exercise appears in the following paragraph.

Since you know something about optimization and trade-offs, you recognize the prime objective here: to find a degree of restraint that minimizes total cost to society. You can view this trade-off in terms of freedoms. Individually and collectively, citizens must resolve a conflict between two types of freedom: the traditional freedom that relates to belief and behavior versus freedom from abuse, exploitation, bodily harm, or even annihilation in an overpopulated world equipped with underrestrained powers. The big challenge is to develop effective means of controlling these powers without unduly encroaching on our traditional liberties.

YOU DECIDE

What additional restraints on the design and use of automobiles do you foresee? What are some of the issues?

____

____

____

We expect the trend to tighter restrictions on automotive emissions and on safety features to continue. Furthermore, unless substitutes are found for oil as the energy source for vehicles, eventually there are likely to be restrictions on the weight and/or horsepower of automobiles, perhaps even on the number of vehicles a family can own.

One issue concerns freedom of choice and action. Current mandated emission and safety features are controversial indeed. Certainly those many who object on restraint-of-freedom grounds will not be pacified by further restrictions.

A second issue concerns the type of restraint. Some requirements are of the performance type, for example, a limit on contaminants in automobile emissions, a minimum on miles per gallon. Other requirements are of the feature type; these include specification of the type of emission control device automobiles must have, a limit on weight, or a specific safety feature. There is a major difference between the two types of restraints. Support for performance standards derives from logic. Support for "feature" standards derives mainly from tradition.

Legal Restraints on Engineers

Government affects engineering in numerous ways, but surely the most meaningful in day-to-day practice is the accumulation of laws and legal precedences within which engineers must operate. You are somewhat familiar with the changing liability attitudes and growing influence of the court. But you may not appreciate to what extent government influences the practice of engineering through legislation.

One example is the Occupational Safety and Health Act (OSHA) of 1970. Basically, it established a host of occupational health and safety regulations plus an ongoing procedure for adopting additional regulations as the need is perceived. Especially those who are engineers designing or supervising the operation of machinery for construction, transportation, manufacturing, electric utilities, agriculture, lumbering, or longshoring, will experience the strong influence of this act. For instance, for designers of cranes, the act specifies safety factors for cables, shielding requirements for moving parts, requirements for steps, ladders, and hand-holds, limits on noise, and so forth.

Another illustration is the 1970 Clean Air Act, with which you are probably generally familiar. You can visualize what it means for engineers. Then there are the Consumer Product Safety Act, Noise Control Act, Interstate Commerce Act, Federal Water Pollution Control Act, and Communications Act, just to name a few.

THINK IT THROUGH

Can you think of any ways that legislation such as that just outlined establishes restraints on the practice of engineering?

Some answers to this exercise appear in the following paragraph.

Legislation such as this directly establishes restraints on the practice of engineering through codes, performance standards, operating regulations, limits on noise, limits on discharges into the air and water, and even through requirements for certain procedural matters concerning contracts and construction permits. But that's not all. Most such legislation also creates regulatory agencies like the Occupational Safety and Health Administration, the Environmental Protection Agency, and the Consumer Product Safety Commission. These agencies generally have enforcement, interpretive, and regulation-setting functions that tend to make them more influential than the original legislation.

Thus, the federal government's heavy hand is felt by engineers through its type I, II, and III decisions. In type III decisions, state governments also play a major role. Considering all the legislation, court-established precedences, and regulatory agencies at federal and lower levels, government's influence over engineering is considerable. But remember, also, that the influence is two-way; what engineers do affects government (generally indirectly), just as what government does affects what engineers do.

SELF QUIZ

1. List the six basic things that are examined in technology assessment.

2. Why is it important for government officials to consider technological proposals and projects in a total context of technology rather than simply as individual projects?

3. List three ways in which technology assessment is future oriented.

4. List two developments in recent years that indicate some increase in governmental capability or activity in the area of technology assessment.

5. Define the term *type III decision*.

6. List two important reasons why the timing of legal restraints on the application of technological capabilities is important.

7. List two important ways in which technology assessment can help the timing of legislation to restrain the application of new technological capabilities.

____

____

____

____

8. List two ways in which legal restraints have been imposed on engineers.

____

____

____

____

1. The six basic things examined in technology assessment are:

- Appropriateness of the proposed design
- Possible alternatives
- Benefits of the proposed design
- Costs of the proposed design, including environmental and third-party costs
- Other effects on the environment and affected populations
- Necessary restraints

2. If technological projects are considered in isolation from the total technological context, we will continue the practice of solving old problems by creating new ones.

3. Technology assessment is future oriented in these three ways:

- It predicts wants and needs.
- It anticipates the need for restricting legislation.
- It sounds early warnings of impending hazards.

4. Two developments in recent years that indicate an increase in governmental capability or activity in the area of technology assessment are:

- The National Environmental Policy Act of 1969
- Establishment of Congress's adjunct Office of Technology Assessment.

5. The term type III decision refers to any decision regarding the amount and kind of re-

straint to be applied to technology. This type of decision also includes questions of what kind of restraint is best in a given case.

6. The timing of legal restraint on the application of new technological capabilities is important because:

- If the legislation comes after the technology is being applied, the users consider the application a precedent and the freedom to apply it to be a right established by usage.

- Major companies will have invested great sums of money in developing the capability to apply the technology and will have very strong vested interests in maintaining the capability.

7. Technology assessment can help improve the timing of legislation to restrain technology by

- Helping predict the total costs of applying the new technological capability, thereby influencing decisions regarding its application and enabling legislation decisions regarding its application and enabling legislation to assign the true costs to those benefiting from the application

- Giving advance warning of hazards and thus enabling sound restraining legislation to be passed before any major investments have been made and before any damage has been done by applying the technology

8. Legislative restraints have been imposed on engineers by the Occupational Health and Safety Act of 1970, the Clean Air of 1970, the Consumer Product Safety Act, the Noise Control Act, the Interstate Commerce Act, the Federal Water Pollution Control Act, the Communications Act, and so on. Such legislation establishes restraints by means of production codes, performance standards, various limits, and even restrictions regarding certain procedural and contractual matters. Such legislation also establishes regulatory agencies that directly impose restraints and sanctions regarding the practice or engineering and related occupations.

FINAL EXAMINATION

Multiple-Choice Test Questions

For each question, there *may* be *one* lettered item which makes an *incorrect* statement. If such is the case, cross out the incorrect item. If there is no incorrect item, circle the letter opposite "None of the above is incorrect." Thus for each question, *one incorrect phrase will be crossed out or one letter will be circled to indicate there is no incorrect phrase.*

1. Some major skills engineers should acquire to practice their profession effectively are in
   a. Design.
   b. Optimization.
   c. Modeling.
   d. Communication (graphical and verbal).
   e. None of the above is incorrect.

2. Skills that engineers should acquire also include
   a. Computer programming.
   b. Experimentation.
   c. Utilizing of information resources.
   d. Working effectively with people.
   e. None of the above is incorrect.

3. The design cycle includes
   a. Periodic monitoring of solutions while they are being used.
   b. Obtaining feedback on effectiveness of solutions.
   c. Gaining acceptance of solutions by clients, supervisors, etc.
   d. The design process.
   e. None of the above is incorrect.

4. Methods for adding system to the search for alternative solutions include
   a. Brainstorming.
   b. Devoting your creative efforts to each solution variable, one at a time.
   c. Use of check lists.
   d. Use of alternatives trees.
   e. None of the above is incorrect.

5. In problem formulation, an engineer
   a. Tries to avoid getting into details.
   b. Identifies the transformation desired.
   c. Should seek as broad a formulation as the organizational boundaries and economics of the situation will permit.
   d. Should consider alternative formulations.
   e. None of the above is incorrect.

6. Decisions concerning what is good design from the point of view of operability (usability) can be based on
   a. Monte Carlo analysis.
   b. Experimentation.
   c. Common sense.
   d. Advice of consultants.
   e. None of the above is incorrect.

7. In evaluating alternative designs, engineers should consider
   a. Indirect as well as direct effects.
   b. Unquantifiable as well as quantifiable consequences.
   c. Guidelines imposed by the American Institute of Engineers.
   d. Consequences for users and nonusers.
   e. Long term as well as short term consequences.
   f. None of the above is incorrect.

8. The process of design
   a. Is the central activity of engineering.
   b. Is not recommended for non-technical problems.
   c. Is the process through which the engineer's knowledge, skills, and attitudes are brought to bear on problems.
   d. Is the general procedure for solving an engineering problem.
   e. None of the above is incorrect.

9. The factors that determine one's inventiveness are
   a. The effort exerted.
   b. Aptitude.
   c. Attitude.
   d. Knowledge.
   e. The method employed.
   f. None of the above is incorrect.

10. As defined in this learning program, a restriction
    a. Is a basis of preference among acceptable alternatives.
    b. Is a "must" characteristic of a candidate solution.
    c. Can also be viewed as a limit on values a solution variable can assume.
    d. Sometimes fixes a solution characteristic.
    e. None of the above is incorrect.

11. Solution variables
    a. Are what you as the designer are out to maximize (or minimize if it is a cost).
    b. Are independent variables.
    c. Are what the designer manipulates in order to create the best overall solution.
    d. Are the respects in which alternative solutions to a problem can differ.
    e. None of the above is incorrect.

12. Concerning problem formulation:
    a. Formulate as broadly as the problem situation will permit.
    b. It is problem definition in detail.
    c. It is the first stage of what amounts to a two-stage problem definition.
    d. There is no such thing as *the* correct formulation of a given problem, but there are some that are more profitable than others.
    e. None of the above is incorrect.

13. Digital simulation
    a. Is appealing because there is no error associated with the results.
    b. Has been applied to the operation of campus parking lots.
    c. Involves a digital model.
    d. None of the above is incorrect.

14. Concerning models and their contrasting roles in science and engineering:
    a. The engineer ordinarily views a model as a means to an end.
    b. Ordinarily, in science, a model *is* the end.
    c. Engineers are prone to devote more attention to developing and refining their models than scientists are.
    d. The history of science is essentially the history of man's attempts to reduce the discrepancies between models and the natural phenomena they represent.
    e. None of the above is incorrect.

15. Assumptions in modeling
    a. Must be avoided.
    b. *Are* the model.
    c. Are nothing to get nervous about as long as the model predicts satisfactorily.
    d. None of the above is incorrect.

16. The random number 0-99 employed in certain Monte Carlo techniques
    a. Can be generated by computer.
    b. Will result in an approximately flat and straight line, if 1000's of such numbers are plotted in frequency histogram form.
    c. Can be generated by a spinner on a pie chart with 0 through 99 marked off evenly around its circumference.
    d. Can be generated by throwing ordinary dice.
    e. None of the above is incorrect.

17. Simulation
    a. Cannot be employed when there is an element of chance involved in the behavior of the system simulated.
    b. Involves experimentation.
    c. Is a form of modeling.
    d. Can be performed on iconic models.
    e. None of the above is incorrect.

18. To predict the performance of alternative solutions, engineers employ
   a. Judgment.
   b. Mathematical modeling.
   c. Melbourne's transform analysis.
   d. Simulation.
   e. None of the above is incorrect.

19. An acceptable Monte Carlo method is
   a. The pie chart and spinner.
   b. A table of random numbers.
   c. An appropriate conversion table along with a source of random numbers.
   d. The values-on-slips-of-paper routine.
   e. None of the above is incorrect.

20. Engineers make use of models for
   a. Communication.
   b. Prediction.
   c. Aids to thought.
   d. Sub-solution calibration.
   e. Training.
   f. None of the above is incorrect.

21. In the typical optimization situation, the dependent variable
   a. Is also known as the *criterion.*
   b. Reaches a maximum (or minimum) as the independent variable is altered.
   c. Is also referred to as the *manipulated variable.*
   d. Is a function of one or more independent variables.
   e. None of the above is incorrect.

22. Feedback systems are involved in
   a. Learning.
   b. Body balance.
   c. Maintaining a company's inventories within acceptable limits.
   d. Sprinkler systems (employed to extinguish fires in warehouses, departments stores, etc.).
   e. Guiding ballistic missiles.
   f. None of the above is incorrect.

23. Automatic control as defined in this learning program
   a. Is exemplified by the flyball governor.
   b. Is exemplified by the system for positioning a drill ship over the drill hole.
   c. Is exemplified by the familiar thermostat system for temperature control.
   d. Is also referred to as a self-regulating or self-correcting system.
   e. None of the above is incorrect.

24. The insensitive zone of a feedback control system
   a. Should be as narrow as possible.
   b. Should be optimized.
   c. Permits a certain amount of oscillation without causing the control system to respond.
   d. Prevents a feedback system from over-responding to relatively insignificant deviations from the set point.
   e. Is established by reponse thresholds.
   f. None of the above is incorrect.

25. Oscillation
   a. In feedback systems can be prevented if some lag is introduced into the system.
   b. That continues without increasing or diminishing is called *hunting.*
   c. That continues to build in amplitude creates an unstable system.
   d. That dies out is said to be *damped.*
   e. None of the above is incorrect.

26. An insensitive zone
   a. Prevents a control sytem from over-responding to inconsequential changes in the regulated variable.
   b. Is established by action limits.
   c. Permits a certain amount of random variation in the controlled variable without triggering a response.
   d. Eliminates oscillation.
   e. None of the above is incorrect.

27. The goal specified for a feedback control system
   a. Is constantly being adjusted to keep it within the insensitive zone.
   b. Is sometimes referred to as the *set point.*
   c. Is the desired state of affairs.
   d. Is exemplified by the temperature setting of the thermostat.
   e. None of the above is incorrect.

28. The internal, working-memory of a computer
   a. Stores data and program.
   b. Serves as a momentary, "scratchpad" memory during execution of a computer program.
   c. Is suitable for long-term storage of information.
   d. Has addressable storage locations.
   e. None of the above is incorrect.

29. The digital computer is affecting the field of engineering by
   a. Reducing the detailed, routine, repetitive aspects of an engineer's work.
   b. Enabling engineers to avoid some of the grossly over-simplified mathematical models they were forced to use in the past.
   c. Reducing the opportunities for engineers to exercise their creativeness.
   d. Stimulating the application of mathematics.
   e. Enabling engineers to make better use of what is known.
   f. None of the above is incorrect.

30. What the modern computer system makes available to the engineer is
    a. Many persons able to share the same computer,
    b. With virtually instantaneous availability to all,
    c. With a large library of programs available to everyone,
    d. With large auxiliary memories for storage of data,
    e. And the possibility of working in conversational and graphic modes.
    f. None of the above is incorrect.

31. The computer's branching capability
    a. Is its ability to compare letters and/or numbers and follow different series of instructions depending on the results of that comparison.
    b. Refers to its ability to process a given job in batch or peripheral mode.
    c. Is the basis of its ability to handle exceptions (e.g., in payroll calculations, the person who worked more than 40 hours).
    d. Is the basis of the system for checking serial numbers of cars sold to see if they have been reported stolen.
    e. None of the above is incorrect.

32. A significant difference between pre- and post-WW II periods is
    a. The number (and concentration) of people around to be affected by technology.
    b. The federal government's hand in financing and promoting new technologies.
    c. The power (for good and for evil) that technology puts at man's disposal.
    d. Our dependency on technology.
    e. The major role of women as precipitators of technical change.
    f. None of the above is incorrect.

33. An engineer's professional obligations include
    a. Keeping abreast of new developments to sustain his ability to perform as an engineer.
    b. Alertness to and concern about the many ways his solutions *directly* and *indirectly* affect the public.
    c. Making sure that his designs are risk free.
    d. Keeping informed about crucial public issues associated with science and technology.
    e. Aiding the public to understand technology and the major issues surrounding it.
    f. None of the above is incorrect.

34. Technology assessment
    a. Is useful in making Type I decisions.
    b. Is useful in making Type II decisions.
    c. Is useful in making Type III decisions.
    d. Is a capability recently acquired by Congress, in the form of an Office of Technology Assessment.
    e. Has produced rather disappointing results and as a consequence has been talked down in this learning program.
    f. None of the above is incorrect.

35. Examples of Type II technical decisions are
    a. The decision not to build a major dam.
    b. The decision (made in the early sixties) to send a man to the moon.
    c. The decision to build a new freeway bypass around the city.
    d. None of the above is incorrect.

36. An example of a Type III technical decision:
    a. Air pollution laws.
    b. Regulations on the use of radio transmitters.
    c. Motorcycle helmet requirements.
    d. Load limits for trucks on public roads.
    e. None of the above is incorrect.

37. Technology assessment should include
    a. Consideration of reasonable alternatives.
    b. Consideration of the "need."
    c. Consideration of the restraints that may subsequently be necessary.
    d. Investigation of the probable benefits.
    e. None of the above is incorrect.

38. Concerning the interaction between engineering and government:
    a. Engineers affect government indirectly, through their many creations.
    b. Engineers affect government especially through legislation like OSHA (Occupational Safety and Health Act) and numerous liability decisions made by the courts.
    c. Government affects engineering through its Type I, Type II, and Type III decisions.
    d. As an outgrowth of modern technology, there is a proliferation of federal agencies that control and/or promote a technical capability.
    e. None of the above is incorrect.

39. Since 1940
    a. The power of man's technical capabilities (for good and for evil) has increased significantly.
    b. The federal government has assumed less of a role as a promoter and backer of new technologies.
    c. The need for greater "impact awareness" of engineers has increased.
    d. None of the above is incorrect.

40. Examples of properly stated criteria include the following:
    a. "Low cost."
    b. "Reliability."
    c. "Safety."
    d. "Aesthetic appeal."
    e. "Environmental effects."
    f. None of the above is incorrect.

Note: Count 1½ points for each correct answer.
Count 4 points for each correct answer in the following section.
Perfect score = 100 points.

Short-Response Test Questions

Answer each question completely, but as briefly as possible.

41. As a part of a digital simulation to determine how many voting machines to provide for voting centers of different sizes, a Monte Carlo mechanism is needed to sample randomly from the following data.

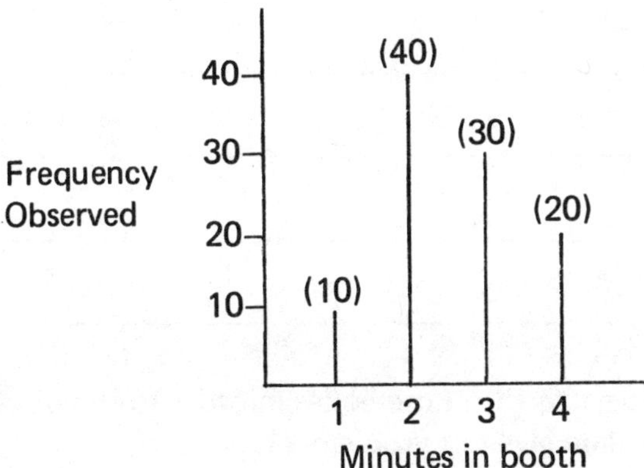

Set up a conversion table for converting random numbers into waiting times, given the information above. Proper labeling of the parts of the table is expected.

42. Given the information in problem 41, prepare a pie chart that might be used as a Monte Carlo device in place of the random number conversion table.

43. Assume that you wanted to avoid the nuisance of preparing a free-spinning pointer for the pie chart you constructed in problem 42. How could you use that pie chart for Monte Carlo purposes without a pointer?

44. Use English or computer language to fill in the boxes in order to show how a computer can randomly select from the voter data given in problem 41.

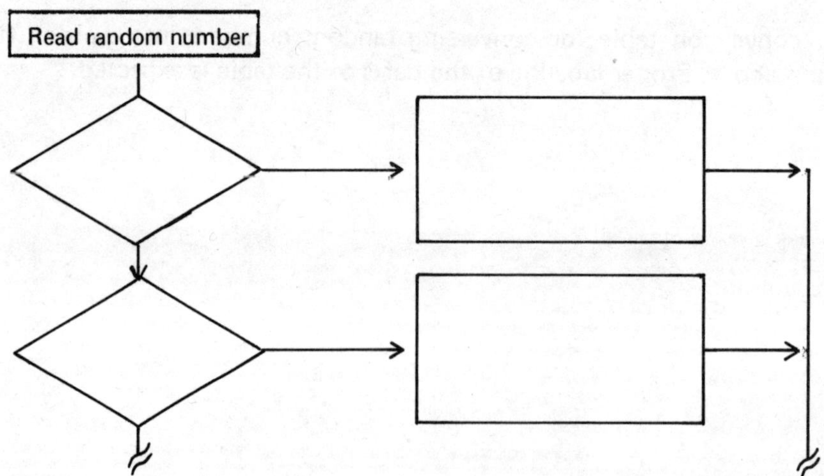

45. Using these axes, illustrate with points what is meant by *global search* and *local search* in optimization procedure.

46. Place the appropriate letter from the terms in List B opposite the most appropriate statement in List A.

Subject: Optimization

List A	List B
One letter per space	
____ Compromising between conflicting criteria	Some letters might be used more than once, some not at all.
____ The dependent variable	a. Analytical optimization
____ Assigning relative weight (importance) to a criterion	b. Trade off
	c. Min-max analysis
____ The independent variable	d. Manipulatable variable
____ Best with respect to a given criterion	f. Value decision
____ One of two *major* methods of locating an optimum	g. Iterative optimization
____ The other major method of locating an optimum	i. Optimization
	m. Optimum
____ An investigation to determine the consequences of not setting a given manipulatable variable at its optimum value	o. Sensitivity analysis
	p. Criterion constraint
	r. Criterion
	s. None of the above

47. Designate which of the following formulations of this problem is **B** (broadest of the formulations given); **N** (narrowest of these formulations).

____Convert gasoline into rotating shaft.

____Convert gasoline into mechanical work.

____Convert fuel into mechanical work.

48. The two-stage problem definition that has been recommended in this course consists of ____________________ and ____

____

49. In the column at the left, place the letter representing the word or expression in the right-hand column that most closely relates to it.

One letter per space	Use each letter once.
____Effector	a. Oscillation
____Response threshold	b. Interdependence
____Goal	c. Insensitive zone
____Feedback	d. Knowledge of actual state of affairs
____Hunting	f. Intended state of affairs
____Feedback control system	g. Thrust propellers on the sides of drill ships

50. For a trestle, if fewer supporting piers are utilized, the pier cost is less but the deck cost is greater. Find the optimum number of piers, given the following:

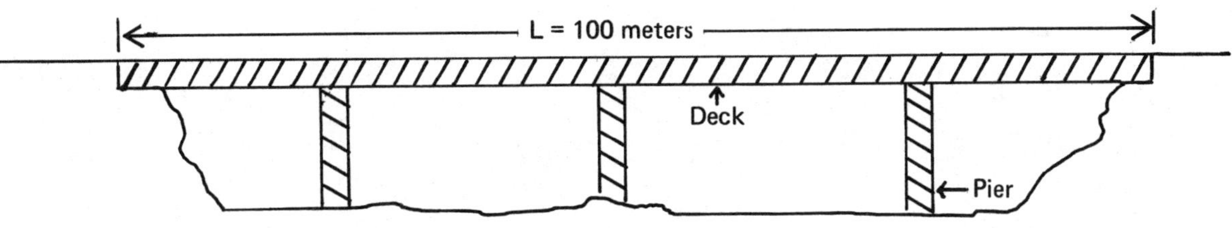

Cost of a pier = \$10,000
N = number of piers

$$\text{Cost of decking} = \frac{\$100{,}000}{N^2}$$

$$\text{Total cost} = N(\$8{,}000) + \frac{\$1{,}000{,}000}{N^2}$$

N	Pier cost	Deck cost	Total cost

Final Examination Answers

1. e
2. e
3. e
4. a
5. e
6. a
7. c
8. b
9. f
10. a
11. a
12. b
13. a
14. a
15. c
16. a
17. d
18. a
19. c
20. b
21. d
22. c
23. f
24. e
25. a
26. d
27. a
28. c
29. c
30. f
31. b
32. e
33. c
34. e
35. b
36. e
37. e
38. b
39. b
40. f

Note: Count 1½ points for each correct answer above.
Count 4 points for each correct answer in the following section.
Perfect score = 100 points.

41. Conversion Table

If random number is	Minutes in booth are		If random number is	Minutes in booth are
0 - 9	1		1 - 10	1
10 - 49	2	OR	11 - 50	2
50 - 79	3		51 - 80	3
80 - 99	4		81 - 100	4

42.

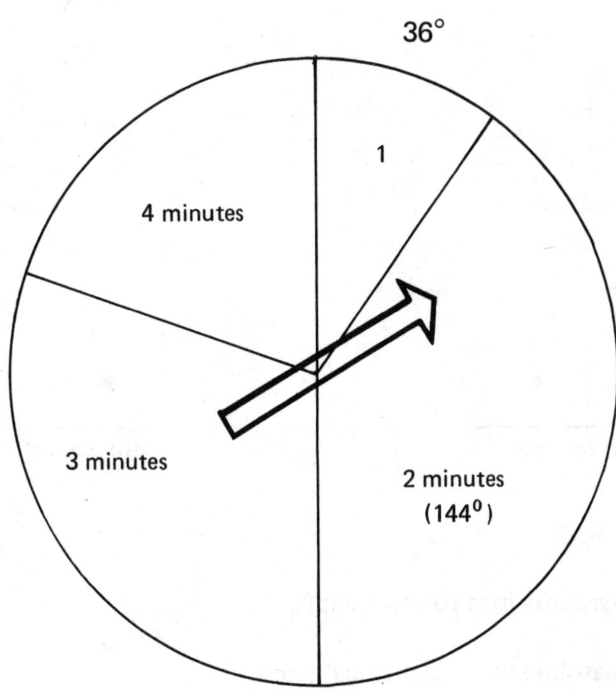

43. Mark off the numbers 0-99 equally spaced on the circumference of the circle. Then use a table of random numbers. Choose a number from that table, see where it falls on the circumference, and then assign that number of minutes.

44.

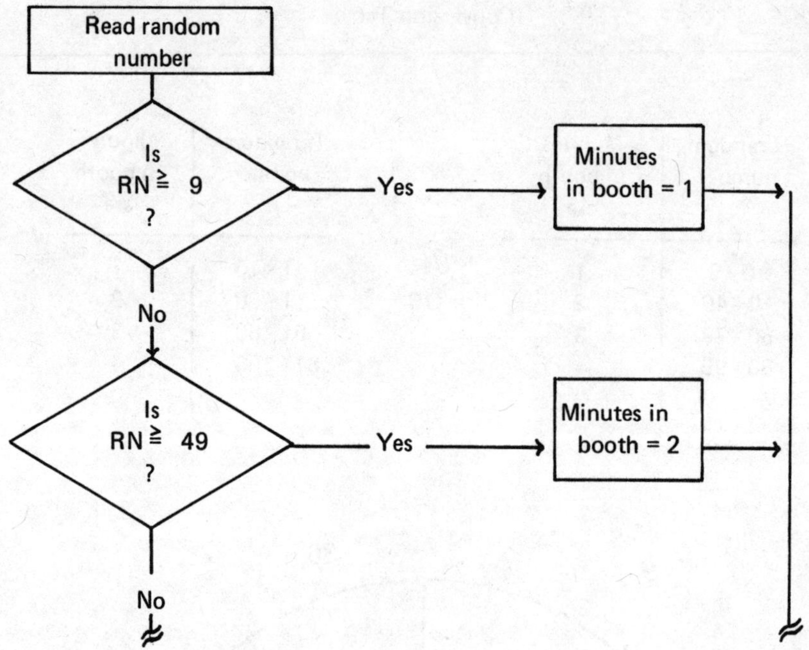

45.

46. b, r, f, d, m, a, g, o

47. N Convert gasoline into rotating shaft.

Convert gasoline into mechanical work.

B Convert fuel into mechanical work.

48. The two-stage problem definition recommended in this course consists of problem formulation and problem analysis.

49. g, c, f, d, a, b

50.

N	Pier cost	Deck cost	Total cost
1	10,000	$100,000	$110,000
2	20,000	25,000	45,000
3	30,000	11,000	41,000
4	40,000	6,250	46,250
5	50,000	4,000	54,000

Optimum is 3 piers.

PICTURE CREDITS

Unit I

1. Chesapeake Bay Bridge—Tunnel District
2. Chesapeake Bay Bridge—Tunnel District
3. Chesapeake Bay Bridge—Tunnel District
4. Chesapeake Bay Bridge—Tunnel District
6. Chesapeake Bay Bridge—Tunnel District
8. Chesapeake Bay Bridge—Tunnel District
11. British Hovercraft Corporation
14. Western Electric Company
15. National Aeronautics and Space Administration
16. Teledyne Ryan Aeronautical Company
17. Teledyne Ryan Aeronautical Company
18. Teledyne Ryan Aeronautical Company
19. Lockheed-California Company
20. Teledyne Ryan Aeronautical Company
21. The Cleveland Bridge and Engineering Company Limited, Great Britain
22. The Cleveland Bridge and Engineering Company Limited, Great Britain
23. British Petroleum Company, Ltd.
24. National Aeronautics and Space Administration
25. Anderson and Anderson Consulting Engineers, and the National Science Foundation
26. Aerovironment, Inc. The device, trade-named Aeroboost, invented by P. B. MacCready and P. B. S. Lissaman
27. a) Cutler-Hammer Inc., b) Lockwood Corporation, Gering, Nebraska

Unit II

15. Photo-archives of the Helsinki City office, copyright by Lehtikuva Oy
18. The Warner and Swasey Company

Unit III

1. Air Products and Chemicals Inc.
4. Dr. J. E. Cermak, Fluid Dynamics and Diffusion Laboratory, Colorado State University
5. United States Navy
6. United States Army
7. The Boeing Aerospace Company
12. Federal Aviation Agency
13. The Boeing Aerospace Company
14. EXXON Corporation
30. Wide World

Unit IV

8. IBM Corporation
11. Lockheed Missiles and Space Company
12. General Motors Research Laboratories
13. General Motors Research Laboratories
14. General Electric Company
15. General Electric Company
16. Burroughs Corporation
20. National Astronomy and Ionosphere Observatory, of the National Astronomy and Ionosphere Center, operated by Cornell University for the National Science Foundation

ISBN 0-471-01702-7